Tech Innovations Inspired by Nature

Don Nardo

ReferencePoint Press®

San Diego, CA

Printed in the United States

For more information, contact:
ReferencePoint Press, Inc.
PO Box 27779
San Diego, CA 92198
www.ReferencePointPress.com

Names: Nardo, Don, 1947- author.
Title: Tech Innovations Inspired by Nature/ by Don Nardo.
Description: San Diego, CA : ReferencePoint Press, Inc., 2024. | Includes bibliographical references and index.
Identifiers: LCCN 2023018257 (print) | ISBN 9781678206086 (library binding) | ISBN 9781678206093 (ebook)
Subjects: LCSH: technology---Juvenile literature

CONTENTS

Attempts to Copy Nature's Designs

Normally, the owners and builders of a recently constructed, supermodern shopping center would be horrified to find that all of the initial reviews of the project mentioned termites. After all, those pesky insects consume wood. And because they can infest and over time weaken and ruin buildings that contain a lot of wood, they have always been linked to structural rot and decay.

The story was very different, however, for the Eastgate Centre in Harare, a fast-growing city in the African nation of Zimbabwe. Its owners were actually proud to have their new mall be associated with termites. This is because the project's architect incorporated natural architectural elements found in the mounds erected by certain species of termite. Those mounds "are constructed like a chimney," environmental activist Jeremy Williams explains. "The mound warms up in the sun during the day, and by nightfall the air inside the mound is warmer than outside. As the outdoor temperature falls, warm air naturally vents out the top of the chimney, which draws cool air in through the many underground tunnels and their exits. This is how termites get fresh air into their homes, and keep them at a stable temperature."[1]

The Eastgate shopping center employs that same method. Its architect, Mick Pearce, set out to mimic termite architecture in the building's cooling system. Exactly like what happens in some termite colonies, air in the lowest level of the Eastgate Centre naturally flows upward. In Pearce's building the air moves through rows of brick chimneys to the roof, where the air vents, or escapes.

Taking advantage of that setup, a series of tunnels running beneath the structure draw in cool night air, trapping it. Later, during the daylight hours, strategically placed pipes carry that cool air upward, a little at a time, throughout the enormous building. Thanks to this natural cooling system, Eastgate has no human-made air-conditioning equipment. As a result, a lot of money was saved in initial construction costs, plus operational costs will be cheaper throughout the center's lifetime.

Aiming for a Balanced Extension of Nature

Pearce's brilliant use of a physical concept found in nature, specifically in termite mounds, is a vivid example of what scientists have come to call biomimicry. The Biomimicry Institute in Missoula, Montana, defines it as "a practice that learns from and mimics the strategies found in nature to solve human design challenges."[2] In a very real sense, scientists at the institute say, over the course of billions of years, nature has undergone a vast process of experimentation and trial and error. A tiny bit at a time, and in numerous clever ways, nature has made various animal and plant species increasingly efficient. Biomimicry is a human attempt to take full advantage of that efficiency.

Put another way, biomimicry allows people to learn from and apply nature's handiwork to fulfill human needs. Moreover, it attains that goal in ways that allow human civilization to be a sort of balanced, beneficial extension of nature rather than a force that works against and damages natural processes. By using biomimicry, a spokesperson for the Biomimicry Institute states,

> we have the innate potential to create a world that is harmonious, regenerative, and supports the well-being of all life. As a young species, we are fortunate enough to have

guidance to help us rise to this challenge. Our planet-mates, the fantastic meshwork of plants, animals, and microbes, have done billions of years of research and development. By infusing biological strategies in design, we can transform sustainable innovation. . . . The goal is to create products, processes, and systems—new ways of living—that solve our greatest design challenges sustainably and in solidarity with all life on earth.[3]

Not a New Idea

The idea of learning from nature is far from new. A few insightful inventors and other innovators in prior ages recognized, for instance, that some animals can perform certain acts better than people could. Furthermore, a few animals are adept at amazing feats that humans cannot do at all. Perhaps the most obvious example of the latter is the act of flying. Birds, bats, and some insects can fly, whereas people lack that ability. One way that humans can fly, therefore, is to closely observe a bird flying and find some way to copy it.

The most famous early example of trying to mimic bird flight consisted of the attempts by Italian inventor Leonardo da Vinci to build a flying machine. His detailed notes on bird flight and how it might be imitated are contained in a document titled *Codex on the Flight of Birds*, completed in 1506. As architect Christopher J. Botham explains, da Vinci's flying machine consisted of a person wearing a harness with artificial wings attached. "The two wings would operate through a system of ropes and pulleys, with the operator's legs providing power by working two small stirrups at the opposite end," he says. "Pushing down on one stirrup caused the wings to move down, while pushing down on the other stirrup caused the wings to move up. Do this over and over, and the wings flap up and down like a bird."[4]

"We have the innate potential to create a world that is harmonious, regenerative, and supports the well-being of all life."[3]

—Biomimicry Institute

Termite mounds, similar to this one in Zimbabwe's Hwange National Park, have stable interior temperatures because of their structure. This served as an architectural model for the Eastgate Centre in the country's capital city of Harare.

Da Vinci's attempt to make a human fly ultimately failed. In large part this was because he wrongly assumed that a person's muscle power alone would be sufficient to lift his or her body into the air. Later pioneers of flight, including the Wright brothers, who also studied birds, correctly realized that the key to flight is airflow, not bodily strength. Making air flow over the wings, they showed, produces enough power to lift a bird, or an airplane, upward.

Hope for a Better Future?

Considering how human flight—achieved by copying nature—has changed the world, the tremendous potential value of biomimicry becomes clear. The world and human society have many large-scale problems to solve. And many scientists are convinced that biomimicry will contribute to solving a good many of them. For instance, the Biomimicry Institute points out, many people

> are losing hope for solving the climate crisis and its many negative effects on ecosystems across the world. Biomimicry gives us hope, because we know the solutions are here, accessible, and validated by the many species contributing to their ecosystems today. By using nature as our mentor, we experience the powerful healing effects [that might come from] connecting to the natural world.[5]

CHAPTER ONE

Engineering and Construction

Ever since humans stopped being hunter-gatherers and settled down in communities, century after century and right up to the present day they have relentlessly built things. Houses, ships, dams, cars, skyscrapers, computers, and entire cities have been built. "Throughout history," says science writer Janine M. Benyus, "our progress as a people has been date-stamped by the types of materials we used—the Stone Age, the Bronze Age, the Iron Age, the Plastic Age, and now, some would say, the Age of Silicon."[6]

One unfortunate truism emerges from civilization's ever ongoing flurry of architecture, engineering, and construction. Namely, no matter which building materials are employed, after the various structures have served their purposes, the materials that remain become piles of waste that must be dealt with in one way or another. A spokesperson for the UK-based biomanufacturing company Biohm says:

> Global construction waste is a major sustainability challenge. It is estimated that construction contributes to [a] third of the world's overall waste and it is also a major contributor to carbon dioxide emissions. . . . [Moreover] the pollution caused by the

construction industry does not seem to be slowing down. According to the World Economic Forum, the waste produced by the construction industry is likely to reach 2.2 billion [tons] per year by 2025.[7]

Hoping to deal with this construction waste problem, Biohm and some other companies have begun to develop new building materials that are biodegradable. That is, when they are no longer in use, nature will break them down into their basic elements, which can then be used to create a host of other products. What makes these building materials particularly striking is that they imitate or utilize elements of nature itself.

One biomimetic example is Biohm's new insulation panels, made of mycelium. The latter is a manufactured version of the complex network of protein filaments within underground fungi. The mycelium panels are strong, durable, and extremely flexible. Equally important, the company points out, each panel "can be safely composted or reused at the end of its life."[8] Thus, mycelium and similar new substances inspired by nature are leading the way toward more efficient ways to dispose of old construction materials.

Copying Small-Scale Natural Strategies

The products emerging from innovative companies like Biohm are only a few of the practical ways that humans are trying to learn from the natural world. Nature has numerous hidden strategies for accomplishing a wide array of mechanical tasks. And modern engineers and builders hope to imitate as many of those strategies as possible.

Some of the tasks that nature found clever ways to accomplish are quite small scale and initially were not very obvious to most people. One now-famous example of such a small-scale engineering strategy is one of several ways that nature has found to make one object cling to another. The first known trained engineer to notice this strategy was Switzerland's George de Mestral. In 1941, while on a hunting trip in the wilds of the Alps, he

The idea for the hook-and-loop design of Velcro (inset) grew out of a hunting trip in the Swiss Alps. The annoying burdock burrs that stuck to the inventor's pants and his dog's fur sparked the idea for the now-ubiquitous fastener.

became annoyed that a number of burdock burrs kept sticking to his pants and his dog's fur.

De Mestral felt compelled to find out what caused this phenomenon. When he looked at the burrs through a microscope, he saw that they were covered with hundreds of tiny hooks, which grabbed hold of most materials the burr touched. This discovery led de Mestral to invent Velcro, which ever after has been used as an effective fastener.

The ability of one surface to cling to another is not the only small-scale engineering strategy that nature worked out over the course of time and that humans eventually adopted. Another example consists of the briar, a natural tool for creating a barrier that keeps one kind of living thing safe from competitors or invaders. Before humans thought to copy the briar, they employed fences and walls to keep intruders out. However, constructing walls and fences from rocks, boulders, branches, or wood was, and still is, time consuming and at times expensive.

All that changed dramatically in the late 1800s when inventors Michael Kelly and Joseph Glidden introduced a metal mimic of the briar—barbed wire. According to inventor Jay Harman:

> Barbed wire has had as much importance in the settlement of the American West, Australia, and other large land areas as guns, the railroads, and the telegraph. . . . The impact of barbed wire on migratory animals and indigenous, nomadic peoples, such as Australian Aboriginals, who had wandered freely for fifty thousand years, or Native Americans, who called it "devil's rope," was devastating.[9]

Concrete That Does Not Add to Climate Change

Today human inventors continue to imitate nature's engineering strategies, including several adaptations used in large-scale construction. One of the most important examples is an extra-strong version of concrete copied from the manner in which coral reefs build up their extremely hard skeletons. Concrete—made from sand or crushed rock and mixed with limestone cement and water—is "the world's most widely used human-made material," science writer Kristy Hamilton points out. "Concrete is under our wheels on highways as we drive bumper to bumper to our jobs, homes and dreams. It is the bones of our cities, used in everything from buildings and bridges to pipelines, drains, and dams. It is both a lifeless building block and a lifeline of humanity."[10] It is used for so many purposes because it is easy to make, is extremely strong, and does not rot or become infested with insects.

> **"Concrete is under our wheels on highways as we drive bumper to bumper to our jobs, homes and dreams."[10]**
>
> —Science writer Kristy Hamilton

But concrete does have a drawback: it emits large quantities of greenhouse gases, which contribute to climate change. In fact, as reported by PBS's news division, a full 8 percent or more of worldwide carbon dioxide (CO_2) emissions each year come from concrete in streets, sidewalks, and buildings. In

The polyps that make up coral may prove useful in counteracting the buildup of greenhouse gases that results when concrete is made. Pictured here are polyps of cave coral, a variety of coral found off the coast of Australia.

addition, the energy expended during the process of making concrete comes from burning coal or oil, which also releases CO_2 into the atmosphere.

To counteract this buildup of greenhouse gases connected to concrete, scientists have been looking closely at coral. Consisting of countless tiny living creatures called polyps, coral slowly forms large masses, called reefs, that are extremely hard. During their life cycles, the coral polyps absorb CO_2 that naturally exists in seawater. And that CO_2 is permanently trapped, becoming part of the structure of the reef.

Some companies are now perfecting ways to collect the CO_2 released by burning coal or oil during the making of concrete. Special machines capture the gas and turn it into a highly concentrated form. Then they inject it into the concrete itself, so that the CO_2 becomes part of its solid mass, mimicking the way coral traps CO_2.

Emulating Bees' Amazing Homes

Modern engineers and architects have copied nature not only in the area of construction materials but also in the shapes and designs of various living and working spaces. One of nature's most successful architectural designs, the conical beehive, has been popular with architects and builders for many centuries. "Bees are amazing creatures," says Australian architect Nina Tory-Henderson.

> Their hives are fascinating structures, and it's no wonder that humans have been inspired by them when designing buildings. Bee hives are typically round or oval in shape, with multiple levels. The honeycomb cells within the hive are also round and arranged in a spiral pattern. This allows the bees to move easily from one cell to another. . . . Humans have used the bee hive as inspiration for everything from office buildings to apartments. The round shape of the hive is efficient and space-saving, and the spiral pattern of the comb can be replicated in staircases and hallways.[11]

A Leg Bone Inspires a Famous Landmark

Roughly 7 million people, mostly tourists from foreign lands, visit France's world-famous Eiffel Tower annually. What many of those visitors do not know is that the tower's designer, French architect Gustave Eiffel, based his iconic structure on the longest and largest bone in the human body—the femur. The femur provides support for the upper leg. Biomimicry Institute writer Adrian Johansen explains that Eiffel mimicked the femur because he realized that the building would need support to withstand many harsh conditions. These included high wind gusts, periodic bouts of rain and snow, and exposure to heat and dryness from sunlight, not to mention the force of gravity constantly pulling on the tower's own enormous bulk. Johansen says Eiffel also knew that "the femur is lightweight but can withstand a lot of pressure—something the Eiffel Tower needed. Inside the human femur is an organized lattice of bone fibers that give it more structure and stability. Anyone who has ever seen a picture of the Eiffel Tower can see a similar woven pattern within the heart of the landmark, providing that same stability, no matter the conditions it faces."

Adrian Johansen, "Looking Within: Design Inspired by Teeth and Bones," Biomimicry Institute, November 22, 2022. https://biomimicry.org.

The Learning Hub at Singapore's Nanyang Technological University is the most famous recent architectural wonder that mimics the beehive. The project, completed in 2015, consists of twelve towers, each shaped like an upside-down cone and all of them interconnected by walkways. The main theme of the huge structure is social interaction between students and teachers and between students and other students. To emphasize that idea, the towers feature a complex mix of, in Tory-Henderson's words, "intimate nooks, plant-filled balconies and large circulation areas that act as social gathering spaces and are seen as equally important informal learning environments."[12]

From the Ocean's Depths to London's Skyline

As it turns out, nature has employed the cone shape in several other instances. One, which is little known to most people, is the Venus flower basket sponge, an organism that dwells in deep, dark ocean waters. According to noted Turkish architect Eda N. Dalgiç, it features a conical, dome-like exoskeleton with a surface that resembles a lattice. The open spaces in the lattice "dissipate forces from powerful streams of water," Dalgiç explains. Also, she says, the fibers making up the lattice "can bend at intersections to absorb stress. This silica skeleton forms the sponge's hollow basket where seawater is filtered for nutrients and flows through it and is directed upwards."[13]

These unique qualities of the Venus sponge caught the attention of British architect Norman Foster, who was hired in 2001 to design a new skyscraper in London. Its official name is the 30 St Mary Axe Building, but it has come to be better known by its nickname, the Gherkin Building. (Gherkins are a type of pickle—and many people feel it looks like a giant pickle.) Foster's goal was to have the structure dissipate the forces of wind in the same way the Venus sponge scatters the impact of ocean water. The Gherkin Build-

> **"The air flows more gently around the [Gherkin Building than] in a traditional skyscraper."[14]**
>
> —Turkish architect Eda N. Dalgiç

ing's conical shape "reduces wind deflections," Dalgiç explains, "thereby creating external pressure differences that operate the natural ventilation system. The air flows more gently around the building, not [head-on] like in a traditional skyscraper. The lattice-like carrier system provides an open floor plan without interior pillars. This opening also helps to let a large amount of natural light into the building."[14]

London's iconic 30 St Mary Axe Building—better known as the Gherkin Building—dissipates the forces of wind much like the deep-sea dwelling Venus flower basket sponge scatters the impact of ocean water.

Saved by the Beak

Airflow is also a key factor in another well-known human engineering accomplishment that was inspired by nature. The Japanese high-speed bullet trains that became world famous in the late twentieth century initially had one notable drawback. Namely, when a train entered a tunnel, strong pressure waves formed, pushed outward, and exited the far end of the tunnel at the speed of sound. That action created a loud sonic boom that disturbed residents of nearby towns. Because completely rebuilding the tunnels was far too expensive, engineers searched for another way to keep the pressure waves from building up in the first place.

The answer came shortly after the researchers considered the idea of changing the shape of the front of the train's lead car. Some of the engineers speculated that nature must have confronted this same problem and solved it. Sure enough, further research revealed that the brightly colored kingfisher bird, common to all the continents except Antarctica, has a highly specialized, streamlined beak. When the bird dives downward at an extremely high velocity, the air flows smoothly around its beak—resulting in no noticeable sound.

Beehive-Shaped Tombs

Conically shaped beehives have inspired architects and builders throughout human history. The most common examples in ancient times were so-called beehive tombs erected by various peoples native to southern Europe, the Middle East, and Asia. The Greeks called such tombs tholoi. A typical Greek tholos consisted of a circular underground chamber made of large stone blocks and having a roof that curved gradually upward, forming a dome, just as in a beehive. A sarcophagus containing the remains of a deceased king, queen, or other noble person rested on the floor beneath the dome. No one knows exactly when and where the first such hive-like tombs appeared. More certain is that the smaller ones were built aboveground. In contrast, the largest ones had to be constructed beneath ground level. This was because their curved roofs were so heavy that the danger of collapse was always imminent. Building them underground greatly reduced that risk because the stones were partially supported by the earth pressing on them from the outside.

When scientists compared the train design to the kingfisher's beak, they saw that the train's front was flat. That shape led to formation of an air pocket, which—when suddenly deflated—made a loud boom. By streamlining the lead car's front to match the bird's beak, the engineers managed to eliminate the sound. Moreover, says journalist Lawson Brandis, "this new model of bullet train not only reduced noise but also reduced air resistance, resulting in the secondary benefits of cost savings and reducing energy consumption."[15]

The human engineers and architects who worked on the bullet trains, along with those who designed the Gherkin Building, got their cues from nature. The more such advances that occur thanks to biomimicry, the clearer it becomes that nature is the ultimate engineer and designer. "After billions of years of research and development," says technology writer Donovan Alexander, nature's failures are now lifeless fossils. And, he states, the highly efficient examples of natural engineering and architectural design that remain for humans to copy "are the secret to survival."[16]

CHAPTER TWO

Medical Applications

Venom, specifically the venom of the Gila monster, is not a substance the average person would associate with weight loss or as a treatment for disease. As it turns out, it is both. Researchers have discovered that the venom of these large, lumbering lizards—found mostly in Arizona and New Mexico—contains a hormone that can help treat human obesity and type 2 diabetes.

Obesity and type 2 diabetes are widespread in the United States—and both pose major risks to health. As of 2023, according to the Centers for Disease Control and Prevention, obesity affected 100.1 million adults, or 41.9 percent, and 14.7 million children, or 19.7 percent. Furthermore, according to the American Diabetes Association, about 37 million Americans—comprising 11 percent of the population—have type 2 diabetes. Both conditions are a major cause of cardiovascular disease.

Medical researchers have spent many years developing treatments for both conditions. In 2021 the US Food and Drug Administration approved a new and promising obesity medicine called semaglutide (which is marketed as Wegovy). "It's phenomenal,"[17] says one researcher. Another describes it as "a big step forward—we finally have something that's reliable and able to produce sustained effects over time."[18]

Semaglutide is an example of biomimicry in action. It involves a human-made copy of a hormone found in Gila monster venom—and its development actually grew out of efforts to find treatments for diabetes. Scientist Daniel

Drucker and his team, working at the University of Toronto, knew that Gila monsters are able to control their blood sugar levels for amazingly long periods. And the hope was that finding out why might help humans control their own blood sugar levels and thereby avoid diabetes.

The first significant breakthrough in the research occurred in 2005, when Drucker and his colleagues created an artificial version of a hormone found in the lizard's venom. From that hormone, the researchers developed a new medicine called GLP-1, which proved effective in treating type 2 diabetes. This achievement led to what medical science writer Julia Belluz describes as "a cascade of additional venom-inspired discoveries."[19] In particular, during tests of the new diabetes medicine, doctors noticed that patients lost significant amounts of weight as a side effect. Further research by several medical teams homed in on that fact, and over the next few years they developed semaglutide.

Bacteria-Fighting Ants

Gila monsters are not the only creatures that have inspired important medical research. Some species of ants, for instance, possess what appear to be built-in disease-fighting abilities—specifically, the ability to fend off harmful bacteria. In the 2010s Clint Penick, a world-renowned expert on ants, decided to devote most of his research time to developing a way to imitate that ability of ants. The hope was to find a way to help humans better fight off infections caused by bacteria.

In this endeavor, the first question that Penick tried to answer was exactly *why* certain ant species are able to avoid bacterial diseases. What he noticed in his studies of these tiny creatures is that when they pass food to one another mouth to mouth, several kinds of bacteria, some harmful and some not, get passed along with the food. Penick observed that the ants are almost always able to neutralize the harmful microbes. This, he found, is because the ants can manufacture antimicrobials, chemical substances that kill the microscopic intruders. Penick felt that if

The venom of the Gila monster (pictured) contains a hormone that can help treat obesity and type 2 diabetes. Scientists were able to imitate the properties of this hormone to create a treatment for these conditions.

he could come up with a way to imitate those antimicrobials, he might be able to make new medicines that would greatly reduce bacterial infections in people.

Particularly puzzling, as well as promising, in Penick's research so far is the discovery of one species of ant that appears to have no antimicrobials at all. And yet its members regularly fend off harmful bacteria. Penick suspects that these ants employ some unknown method to avoid infection. His goal is to eventually find that method and copy it for use in humans.

A Dead Creature Returns to Life

In a similar way, researchers in China and elsewhere are trying to expand human medicine by imitating a quality that enabled a dead jellyfish to quite literally come back to life. This extraordinary occurrence took place in 2015 at China's Xiamen University. A few years before, Jinru He, then a graduate student at that institution, had acquired a baby moon jellyfish. As part of a study of the creature's life and death cycle, he closely observed it in a lab over the course of almost two years until it appeared to die. Based on research done by other Chinese scientists during the preceding decade, the young man suspected that this species of jellyfish might possess a special quality that would actually allow it to return to life. So instead of discarding the creature's body, he placed it in a tank of fresh water and continued the study.

Sure enough, to He's delight, three months later the remnants of the jellyfish began to reconstitute. "Tentacles emerged," writes jellyfish expert Juli Berwald. "A mouth formed. Eventually, perfectly healthy polyps sprang to life from the [the creature's] carcass. Rather than simply die, the jellyfish had seemingly reversed its life cycle."[20] Indeed, it appeared that some of the original creature's long-dead cells had somehow rejuvenated.

Although He and other scientists at the university were not entirely sure how the creature had regenerated, they suspected it was some form of cloning. That is the process by which cells from one organism grow into a youthful replica of that organism. Researchers around the world continue to study moon jellyfish. They hope that they can learn how those creatures rejuvenate so researchers can mimic that process in human cancer cells. Unlike infectious diseases, which are caused by bacteria and other microbes, cancer occurs when some of a person's cells become abnormal and multiply uncontrollably. The hope is that lessons learned from studying moon jellyfish might someday allow doctors to reverse the cellular growth of some forms of cancer and enable the cells to return to normalcy.

Regrowing Lost Body Parts

Just as some jellyfish have long been studied with the process of biomimicry in mind, so too has a species of salamander called the axolotl. Its members are cannibalistic; when they are born they and their parents begin attacking and eating one another. And only a few in each new generation manage to survive. In "such a harsh nursery," remarks science writer Amber Dance, over time they developed the incredible capacity to regrow lost body parts. "If an axolotl loses a limb," she points out, "the appendage will grow back, at just the right size and orientation. Within weeks, the seam between old and new disappears completely. And it's not just legs:

> “If an axolotl loses a limb, the appendage will grow back, at just the right size and orientation.”[21]
>
> —Science writer Amber Dance

Axolotls can regenerate ovary and lung tissue, even parts of the brain and spinal cord.”[21]

In fact, scientists have known for decades that axolotls have this seemingly miraculous ability. For example, studies have shown that an axolotl that has lost a leg grows a layer of a special kind of skin at the spot where the limb was severed. From that spot, a type of tissue called a blastema eventually appears, and soon afterward a new leg sprouts from it. Unfortunately for the researchers, however, although they watched this process unfold on many occasions, the details about how the salamanders’ DNA, cells, and tissues made it happen remained largely a mystery.

That situation began to change in the early 2000s, however, thanks to biologists’ sequencing (revealing the structure of) the DNA of axolotls. That gave the investigators the ability to modify several salamander genes in a lab and perform various other technical experiments with them. The ultimate and exact ways that the creatures regenerate body parts still remains somewhat

Medical researchers are trying to understand the physiological secrets of the axolotl, a salamander that can regrow lost body parts. Researchers hope to one day use this knowledge to help people who have lost limbs.

Land Slug Goo for Heart Surgery

Each year about 2 million people around the globe have open-heart surgery. And sometimes surgeons either purposely or accidentally cut into the heart tissue. Those incisions must be closed in order for the patients to survive. But ordinary means such as sutures are not always effective. In part that is because the heart is constantly beating and the sutures do not expand and contract like the heart tissue does. Hoping to invent a more flexible adhesive for use in heart surgery, in the 2010s researchers at Harvard University's Wyss Institute for Biologically Inspired Engineering began intensely studying land slugs. These creatures move along by secreting a gooey substance they can glide through. The researchers noted that when a slug feels threatened, it secretes more goo. This goo contains a strong adhesive that enables the slug's body to stick to any surface so that a predator cannot dislodge the slug. Moreover, the sticky goo can expand and contract without losing strength. Inspired by this idea, the scientists developed a sticky tape that they used to neatly seal a cut in a pig's heart. Tests on human hearts are ongoing, and the new adhesive may be in wide use by the mid- to late 2020s.

elusive. But biologists feel they are tantalizingly close to learning these amazing secrets of nature.

A breakthrough in this research could revolutionize the science of medicine. When the axolotl's secret is revealed, Dance suggests, medical researchers will be able to develop ways to create a similar regenerative process in humans. Perhaps within a few decades, she goes on, "people, too, might regrow organs or limbs."[22]

Porcupine Quills and Surgery

In the meantime, while science awaits such giant leaps forward, medical researchers are taking advantage of less dramatic, yet still crucial, biomimetic advances in surgical techniques. One of the more important examples consists of key improvements in surgical staples employed to close incisions during operations. Around the globe surgeons perform well over 300 million operations each year. And for decades they have used traditional sutures and staples in the final stage of those procedures. Though reliable, both of those methods have certain drawbacks. Sutures require penetrating the skin with a needle in order to insert the thread that joins two pieces of skin. And despite being faster than sutures, traditional metal staples tear small amounts of tissue as they are applied.

Scientists are trying to create a new generation of surgical staples that mimic the action of porcupine quills. Once embedded in the skin of an attacker, the quills of the North American porcupine (pictured) remain securely in place thanks to hundreds of tiny backward-facing barbs.

In a series of studies carried out beginning in 2012 at various Boston universities and hospitals, Harvard Medical School professor Jeff Karp and some colleagues have tried to create a new generation of surgical staples. Based on the quills of the North American porcupine, when perfected these devices will be far less damaging to tissue. They will also be less painful for patients in the weeks following operations.

The idea for staples that mimic porcupine quills derived from earlier observations by biologists. When porcupines are threatened, they turn to strike the attacker with their tail. The quills on their tail easily detach and penetrate the skin of the animal. Once embedded in flesh, the quill remains in place extremely securely, in some cases for years. The reason for this sturdiness and longevity, the experts noted, is that a quill has hundreds of tiny backward-facing barbs that firmly grip an animal's inner tissue.

Reporting on Karp's work, wildlife video producers Josh Cassidy and Laura Shields explain the researcher's rationale for using porcupine quills as a model for a unique and highly effective wound sealer.

> In contrast to a barbless quill or a surgical staple, which tear the tissue and create gaps that are susceptible to infection, the barbed quill's design means it does minimal damage on the way in, the researchers found. A new type of medical staple that had two barbed tips would require much less effort to place, Karp figures, and the gripping power of the barbs would hold it in position without needing to bend the staple. Karp says he anticipates making the new staples out of biodegradable material so they will fully dissolve over time without having to be removed.[23]

The challenge, Karp says, will be to accurately mimic the barb's shape. "Nature has designs that humans can't achieve yet, at least at large scale,"[24] he says. Even so, he anticipates testing and possible use of the new staples in the near future.

Help from Hippos

Karp's observation of the cleverness and effectiveness of nature's many designs for plants and animals is also reflected in another new advance in medical science. Each year millions of people around the world get painful sunburns. In the United States alone, about 1 million people develop skin cancer each year because

An Arm like an Octopus Tentacle

Artificial arms for people, called prosthetic devices, are extremely difficult to make. This is because the human arm is highly flexible. Along with the hand and its fingers and opposable thumb, an arm can accomplish numerous maneuvers and tasks that most animal appendages cannot. Hoping to create the most effective prosthetic yet devised, in the 2010s University of Washington graduate Kaylene Kau, a prosthetics designer, began studying octopuses. An octopus has eight very flexible limbs, called tentacles, each of which can wind around and tightly grip an object. Based on such a limb, Kau fashioned an arm made up of a long plastic tube divided, from shoulder to hand, into sections of decreasing size. She installed cables linking the sections and a small motor to control the cables. The arm is highly flexible and, like the octopus, can curl around and grip an object. Although Kau's clever creation is still experimental, medical scientists are confident it will eventually be used extensively by people in need of new limbs.

of too much exposure to sunlight. As a result, says Jay Harman, "the market for sunscreen is substantial—$640 million a year and growing." Moreover, of the hundreds of sunscreen products available, "three out of four have been shown not to live up to their claims. In addition, most chemicals applied to the skin can be absorbed straight into the bloodstream, so some scientists are concerned that certain sunscreens may prevent sunburn but introduce toxins that can still cause cancer."[25]

> **"Certain sunscreens may prevent sunburn but introduce toxins that can still cause cancer."[25]**
>
> —Inventor Jay Harman

As it turns out, people who spend a lot of time outside in direct sunlight will in the near future be aided by a large mammal that does not get sunburned. That creature is the hippopotamus. The help from hippos is the result of ongoing studies at Japan's Kyoto Pharmaceutical University, Italy's University of Trieste, and the University of California, Merced. Researchers at these institutions have found that hippo skin is protected by tiny, reddish-colored crystals that are secreted in their sweat. The crystals scatter sunlight, thereby keeping the rays from burning the creatures' hides. The researchers hope to develop artificial crystals similar to the ones in hippo perspiration for use in a new kind of sunscreen. Potentially, they say, such a product would greatly reduce the incidence of sunburns and skin cancer in humans.

Whether it is studying lizard venom to treat obesity and diabetes, learning from salamanders how to regenerate lost tissue, or imitating jellyfish and hippos to develop ways to fight cancer, medical researchers are increasingly turning to biomimicry. It makes sense to learn such things from nature, according to David Gardiner, a biologist at the University of California, Irvine. After all, he says, nature has been tackling problems like disease and loss of limbs far longer than people have. Eventually, Gardiner believes, most medical challenges will be overcome through biomimetic research. He adds, "It's just a matter of time."[26]

CHAPTER THREE

Protective Gear for Soldiers and Athletes

With considerable fanfare, in 2015 the US Army announced that it was officially adopting the Operational Camouflage Pattern (OCP), a mix of colors adorning the fabric of some of its uniforms. In nature, camouflage is a visual system intended to make an animal blend with its surroundings. That way it has a better chance of avoiding predators and surviving. Or if the animal is a predator, it can use camouflage to more effectively lie in wait for its intended prey. Similarly, modern armies employ camouflage to remain hidden from enemy combatants.

The OCP mimics patterns and colors found in nature. A typical forest or countryside features a jumbled mixture of greenish leaves, brownish tree branches, and grayish rocks. And the OCP used in the fabric of soldiers' uniforms copies that mixture of colors.

Yet although the OCP has proved effective to some degree, scientists who study camouflage say that nature itself remains the true master of that art. In particular, they point out, certain sea creatures can hide themselves far more effectively than any human soldier can. Science writer David Szondy explains why, saying,

> The sea can be an extremely savage place, with one animal eating another only to be gobbled up by a third—sometimes while it's still eating the first. It's small wonder then that over hundreds of millions of

years the underwater world has become a giant laboratory of offensive and defensive technologies locked in a perpetual arms race. One of the most effective defenses developed by sea creatures is camouflage and the most adept practitioners are the cephalopods—a class of tentacled mollusks that includes squid, octopi, cuttlefish, and the nautilus.[27]

Highly impressed by this ability of cephalopods, Northeastern University chemist Leila Deravi began studying them in the 2010s. She and her team found that the creatures' remarkable stealth capabilities stem from small, muscular organs called chromatophores. Each of the latter holds many thousands of tiny particles of pigment; the colors include various shades of yellow, orange, brown, and red. If a cephalopod decides to camouflage itself, it quickly retracts a thin membrane that in a mere fraction of a second reveals an appropriate combination of the colored particles.

An octopus changes its colors to blend into its surroundings. Army researchers are trying to mimic this ability so that soldiers might one day be able to rapidly alter their uniform colors for better camouflage.

Working with a team of US Army researchers, Deravi managed to fabricate tiny, colored granules similar to those in cephalopods and inject them into various fibers. She hopes that within a decade or two soldiers will have the ability to make their uniforms rapidly change color. "In other words," Szondy says, "battle fatigues [will] turn a soldier into a combat cuttlefish able to better blend with his surroundings."[28]

> **"Over hundreds of millions of years the underwater world has become a giant laboratory of offensive and defensive technologies."[27]**
>
> —Science writer David Szondy

Soldiers Saved by Sea Turtles?

Another common form of protection for soldiers and other military personnel is body armor. Various forms of it have existed throughout human history, of course, made of materials such as metal and layers of leather or tough fabrics. In the late twentieth and early twenty-first centuries, however, military organizations have increasingly turned to biomimicry for ideas for manufacturing new types of armor. Experiments have been done with the natural armor of armadillos, the scales of some types of fish, the shells of crabs and other crustaceans, and the hides of alligators.

One promising research effort involves the shells and skulls of turtles, especially large sea turtles. One of several scientific investigations of these naturally armored creatures has taken place at the world-famous Woods Hole Oceanographic Institution in Falmouth, Massachusetts. Led by biologist Darlene Ketten, the researchers were originally testing the effects of underwater explosions on dolphins and other sea creatures. (No live creatures were harmed; rather, the tests were done on the bodies of already deceased animals.)

A major unexpected result of these experiments was that the bodies of sea turtles are astoundingly resistant to the powerful force of underwater shock waves. "Basically," Ketten states, "sea

Stronger Airplane and Ship Surfaces

Sharks have inspired innovations ranging from streamlined swimsuits to more productive windmills to improved fuel efficiency in cars. The structure of sharks' skin has also inspired the development of a type of paint that substantially strengthens the outer surfaces of aircraft and warships. Scientists at Germany's Fraunhofer Institute for Manufacturing Engineering and Applied Materials Research announced the paint's development in 2010.

Inspired in part by the dermal denticles on sharks' skin, the researchers created tiny particles similar to the denticles. These microscopic granules have the ability to withstand extreme temperatures and powerful UV radiation. One of the scientists, Volkmar Stenzel, explained to reporters some of the new paint's advantages for aircraft. It is so strong and protective, he said, "that no other layer of material is required. It adds no additional weight and dramatically reduces the cost of stripping a plane every five years (when all the paint is removed and then reapplied). In addition, it can be adhered to complex three-dimensional surfaces without a problem."

Quoted in Jay Harman, *The Shark's Paintbrush: Biomimicry and How Nature Is Inspiring Innovation*. Ashland, OR: White Cloud, 2013, p. 88.

turtles are shielded from shock waves." That prompted her and her colleagues to think that "perhaps there is some [potential here] for human use."[29]

One of the first examples of a benefit for people that came to mind was body armor for soldiers. Examinations of the shells and skulls of the turtle cadavers that had been subjected to the underwater explosions revealed extremely minor damage. Ketten remarked that the bigger turtles looked like submerged tanks, and their skulls closely resembled the helmets worn by many soldiers in World War I. When she studied the turtles, she found that their skulls were thick enough to deflect most of the shock waves from the blasts.

> "It would be very cool to have our soldiers protected by 'turtledom.'"[30]
>
> —Biologist Darlene Ketten

The idea of mimicking sea turtle bone structure to create military body armor had not been part of the original Woods Hole study. But Ketten now clearly saw the potential for saving the lives of many people. "It would be very cool to have our soldiers protected by 'turtledom,'"[30] she says.

Beaver Pelts Inspire Better Wet Suits

Other types of protective gear have also been inspired by nature. Surfers spend hours in ocean water that can be either very warm or very cold, depending on the location. In cold water, surfers usually wear wet suits that help them retain body heat. Most wet suits today are made from neoprene, a stretchy, durable, insulating material. But in the 2000s and 2010s, teams of researchers at the Massachusetts Institute of Technology (MIT) and a few other universities set out to find a material that would be even more effective than neoprene for wet suits.

The scientists took their inspiration from beavers. Beaver fur consists of up to twenty-three thousand individual hairs. Some of the hairs are part of an undercoat that traps air, creating a barrier that keeps moisture out and the creature well insulated. That allows it to spend time in cold water, hop out onto dry land for a while, and then return to the water, all without getting a chill or hypothermia.

Although traditional neoprene wet suits keep surfers warm in cold ocean water, a new material might work even better. It is inspired by beavers, which have an undercoat that traps air and creates a barrier to insulate the animal and keep moisture out.

Anette Hosoi, a professor of mechanical engineering at MIT, led one of the teams that developed the new material, which some companies are now using for wet suits. She later explained that she was "particularly interested in wetsuits for surfing, where the athlete moves frequently between air and water environments. . . . [Surfers] want to be nimble and shed water as quickly as possible when out of the water, but retain the thermal management properties to stay warm when they are submerged."[31] Scientists and athletes alike who have tested the new wet suits agree that they are lighter and less restricting to wear than neoprene versions.

These improvements stem from researchers' success at mimicking the large number of hairs in a beaver's pelt. They began with several blocks of acrylic with a rubbery consistency. Guided by a specially designed software program, a machine studded with ultrathin needles drilled holes into the acrylic blocks. That created empty molds, which the researchers filled with a soft rubberlike substance called polydimethylsiloxane (PDMS). When removed from the blocks, the many thin shafts of PDMS were attached to the surface of a wet suit. Tests confirmed that these artificial hairlike shafts did form air pockets much like those created by beaver fur. In fact, the tests showed that the thicker the fur, the more substantial the air pockets.

Geckos and Better Night Vision Goggles

In recent years the gecko, an amphibian widely known today as a cute talking animal in television car insurance commercials, has become the subject of much biomimetic research by military engineers and other military researchers. Real geckos have large eyes that open particularly wide at night. For that and other reasons, their night vision is some 350 times better than that of humans. (This is essential to geckos because they hunt at night and therefore would not be able to survive without superb night vision.) Moreover, biologists point out that geckos may be one of a handful of animal species that can see color at night. Army engineers are currently studying the creatures to learn how to make better cameras and contact lenses for soldiers, along with advanced goggles that produce night vision of unprecedented power and clarity.

An Idea from Nature's Vast Design Storehouse

Swimmers are also seeing the benefits of biomimicry in their sport. Inspired by an idea from nature's vast design storehouse, in the early 2000s the Speedo swimsuit company set out to make suits for competitive swimmers by copying sharks' skin. Speedo dubbed such outfits Fastskin. This name seemed appropriate, considering that the suits were designed to allow swimmers to move measurably faster in the water.

> **"[Sharks have an] extraordinarily streamlined design, more so than any human-designed object."[32]**
>
> —Inventor Jay Harman

The reason for copying sharks for swimwear, Jay Harman explains, is that those often fearsome sea creatures have an

> extraordinarily streamlined design, more so than any human-designed object in a number of ways. A shark is somewhat bullet-shaped, which reduces its profile [its size and shape as seen by other sea creatures] as it moves through the water. Its fins and tail muscles work in concert to create water vortices [small whirlpools] in their wake, which the shark can, in effect, lean against as it propels itself forward. Finally, it gains drag resistance [the ability to resist the pressure of surrounding water] thanks to its cleverly evolved skin, which is made of tiny vertical scales known as . . . dermal denticles.[32]

Researchers have found that when these denticles make contact with the water flowing over the shark's skin, in a sense they cause that liquid to become a bit choppy. The denticles also push the water backward. This is the same way that a ship's hull pushes water aside as the vessel moves forward through the water. "Imagine," Harman says, "dragging your arm through the water in a bathtub or swimming pool. You can feel the pressure on your arm as the water molecules flow around the obstruction (your arm). What is harder to feel, but is also occurring, is that your arm is dragging water along as well. In a similar way, a ship can drag its own weight in water as it travels."[33]

Researchers found that the little denticles on a shark's skin reduce this effect of drag caused by the creature's movement through the water. Accordingly, Speedo copied that concept by adding hundreds of small, denticle-like ridges to its swimwear. This worked. Competitive swimmers who tried the suits were definitely able to swim faster. The Fastskin suits proved to be *so* successful, in fact, that they became controversial. For various reasons, not all swimmers were able to wear the suits in competition. And for that reason, the officials who oversee the sport decided that those swimmers who did wear the suits had an unfair advantage. Thus, the officials banned the use of Speedo's Fastskins in some professional competitions in 2010. (Nevertheless, many swimmers still wear them outside of such competitions.)

Avoiding Brain Injuries

Football is another sport that stands to benefit someday from technology that imitates nature. In recent years the sport was rocked by many studies showing high levels of brain injuries, including concussions, in players. The studies also indicated that these injuries are caused mainly by the repeated, highly jarring tackles and other hits that players suffer during games. Making matters worse, doctors say that many of the brain injuries lead to detrimental effects such as memory loss, depression, and loss of motor skills.

The players do wear helmets, of course. And several times between 2002 and 2022, various improvements were made in hopes of making the helmets safer. However, critics, including many scientists and doctors, say traditional helmets still do not do enough to prevent brain injuries. As a result, researchers in labs in several US universities have been looking for new and better designs and materials for football helmets.

When rams butt heads, their skulls absorb the impact and disperse the resulting shock waves. Scientists are building on this knowledge to design helmets that provide greater protection to football players who experience frequent hits to the head.

The overall goal is to find ways of fashioning helmets that will significantly reduce the harmful effects of the hits players frequently take. One of the studies, conducted primarily in 2016 and 2017 at Mississippi State University, examined the skulls and brains of woodpeckers and rams. Both creatures regularly sustain violent shocks to the head; woodpeckers pound away at tree trunks, and rams often butt heads with rivals. What interests scientists is why these animals do not suffer permanent brain damage from such acts.

First, the study showed that the composition of a woodpecker's beak is unique. It allows the birds to absorb shocks up to ten times greater than those most football players sustain. Also, the study found, a ram's skull does more than absorb the initial impact of a butt to the head; that skull also channels and disperses the shock waves such a butt produces. In contrast, said one of the Mississippi researchers, engineering professor Mark Horstemeyer, traditional football helmets do nothing to disperse those highly damaging shock waves.

This knowledge allowed the researchers to design a far more effective helmet. "Using information from our research," Horstemeyer explains, "we replaced traditional helmet lining with a foam made of microstructures similar to those found in rams' horns. Tests show our helmet is three times more likely to reduce concussions than helmets worn today."[34]

Hoping to Make the Game Safer

Horstemeyer and his colleagues hope that helmets like those they designed will become standard throughout the sport within a few years. That will undoubtedly revolutionize the game by making it far safer to play, they say. University of South Carolina scientist Dan Speiser agrees. Animals like woodpeckers and rams "seem to risk brain damage all day, every day,"[35] he points out. He looks forward to the day when athletes in rough sports will take full advantage of the tricks by which certain animals avoid brain injuries. Likewise, whether for improved camouflage, tougher body armor, or suits that keep surfers warm, scientists continue to expand their investigations of the ways nature protects its own.

Energy

"Could biomimicry revolutionise renewable energy?"[36] That is the headline of an article on the British power industry news website, Power Technology. The answer appears to be a resounding yes. The natural world has inspired many new technologies—in both renewable energy and nonrenewable energy—and in how people use energy on a daily basis.

Insect Eyes and Solar Cells

Solar power is gaining as a source of renewable energy. That is largely the result of new and better ways of capturing the sun's rays and transforming them into a usable power source. Ideas and efforts to improve this technology have come from many places—including the natural world. One such idea has come from the compound eyes of flies. "We were inspired by the compound eye of the fly, which consists of hundreds of tiny segmented eyes," says California's Stanford University scientist Reinhold Dauskardt. "It has a beautiful honeycomb shape with built-in redundancy: If you lose one segment, hundreds of others will operate. Each segment is very fragile, but it's shielded by a scaffold wall around it."[37]

Dauskardt and his colleagues at Stanford University have been searching for ways to use more efficient and less expensive materials for solar panels. Traditional solar panels are made up of silicon cells. Although it is efficient, silicon cell production is expensive. So Dauskardt's team turned to a different material called perovskite, which works as well as silicon cells but is much cheaper to make. One

"[An insect's eye] has a beautiful honeycomb shape with built-in redundancy."[37]

—Stanford University scientist Reinhold Dauskardt

significant problem plagued the team, however. "Perovskites are the most fragile materials ever tested in the history of our lab," one of the researchers remarks. "This fragility is related to the brittle, salt-like crystal structure of perovskite, which has mechanical properties similar to table salt."[38]

The honeycomb structure of the fly eye offered the team a way to make a flimsy perovskite solar cell a good deal sturdier and more resilient. The researchers cut up the perovskite into hundreds of mini cells, each measuring only a fraction of an inch. They then glued these tiny cells to a honeycomb-shaped scaffold made of a common epoxy resin. Tests showed that because the individual units of perovskite were attached to a strong supportive structure, they were stronger and far fewer became damaged through regular use.

Solar panels made from perovskite cells are not yet in wide use, mainly because researchers are still working out some minor difficulties in their manufacture. Thanks to innovations modeled

The honeycomb structure of fly eyes (seen here on each side of a highly magnified image of a fruit fly) gave researchers an idea for strengthening new materials being used for solar cells.

Plant Leaves and Future Solar Panels

The structure of plant leaves might offer ideas for developing cheaper and more efficient solar panels in the future. Jefferson Lam, a PhD student at the Australian Academy of Technological Sciences and Engineering, is studying the possible application of plant leaf structure to solar panel design. In a March 2023 interview, Lam explained this idea:

> Plant leaves and solar panels are really similar. They're both devices that convert solar light rays into usable energy. Leaves have evolved and optimised over millions of years to become efficient harvesters of this light to feed and nourish plant life while also becoming durable enough to withstand the elements in all environments. By probing into the surface structure of plant leaves and discovering how they handle light so efficiently while also being weather resistant, we can begin to replicate similar structures onto solar panels. . . . I hope that my research will be able to guide the development of new, improved architectures of solar panels so that they are more efficient, affordable, lightweight and durable.

Quoted in Monash University, "Plant Power—How Leaves Are Inspiring the Next Generation of Solar Panels," March 3, 2023. www.monash.edu.

on fly eyes, however, the new material offers enormous promise for improving solar power technology. "The ease with which these higher performances are reached in this new material are almost stupefying," says Tonio Buonassisi, professor of mechanical engineering at MIT. "That's one of the reasons it's so exciting."[39]

Help for Wind Turbines from Whales?

Another form of renewable energy—wind power—has also been influenced by nature. Hugely promising for helping meet global energy needs are many large wind farms that have sprouted up around the world since the turn of the twenty-first century. The largest of these energy-producing complexes, the Gansu Wind Farm in northern China, features over seven thousand turbines and emits enough electricity to power some 14 million homes. The biggest US wind farm—the Alta Wind Energy Center in California—puts out roughly three-quarters as much energy as the Gansu Wind Farm does each year.

When these and most other existing wind farms were constructed, they all suffered from a drawback that kept them from reaching their full potential. Namely, conventional wind-turbine rotor blades are sometimes damaged by sudden, powerful gusts of wind. "When a typical blade sees high variation and gusts of wind speeds," Jay Harman writes, "it can go in and out of stall conditions,"[40] meaning it can slow down or even stop and require a restart.

To try to counter that adverse effect, a number of researchers have suggested employing a kind of biomimicry based on the physiology of whales. In response, a Toronto-based energy company, WhalePower, launched a major study of tubercles—tiny bumps or ridges on a humpback whale's flipper. These bumps reduce water-induced drag and increase lift, allowing the creature to swim more efficiently.

Realizing that a wind-turbine blade is shaped very similarly to a whale flipper, the WhalePower researchers applied small bumps resembling tubercles to the edges of such a blade. Harman describes why the experiment was successful, saying:

> Like a whale using its fins effectively to steeply curve around as it swims, WhalePower's approach of adding bumps to the front of the turbine blades helps minimize the impact of variations in wind speed. The channeling of air by the bumps results in less strain on the equipment and expands the range of wind speeds that can generate power, even in gusty conditions.[41]

Owls to the Rescue

Another common downside of the conventional turbines on wind farms is that these machines make quite a racket. Indeed, during the initial years of their operation, the tall, ungainly turbines were notorious for their high noise levels. This was a problem for people living nearby. It also seemed to have a negative effect on animals in the area. Researchers gathered evidence suggesting

A great horned owl flies in the Arizona desert. Researchers believe they can reduce wind turbine noise by developing blades that mimic the structural configuration of owl wings.

that the clamor created by a turbine's spinning blades made it difficult for some animals to communicate and interfered with some species' mating habits.

The problem appeared to be the turbine blades themselves, so researchers at universities around the world set out to develop a new, quieter blade. Researchers at China's Xi'an Jiaotong University were among the first teams to make progress in this endeavor. For inspiration, the team's members turned to the wings of owls, widely renowned among all bird species for making the least amount of noise while flying. "Nocturnal owls produce about 18 decibels less noise than other birds at similar flight speeds due to their unique wing configuration," states one of the study's leaders, Xiaomin Liu. "So the study of the wing edge configuration during owl flight is of great significance."[42]

> "Nocturnal owls produce about 18 decibels less noise than other birds at similar flight speeds due to their unique wing configuration."[42]
>
> —Chinese scientist Xiaomin Liu

Trying to Mimic Water-Collecting Beetles

One way that humans expend a lot of energy is by trying to obtain fresh water in dry regions that get little or no rain. The scientists at NBD Nano, a new technology research company located in Lexington, Massachusetts, hope to solve this problem by turning to biomimicry. More specifically, they are studying the Namib desert beetle, native to extremely dry regions in Africa. Some water *can* be found each day in such areas in the form of morning fog or dew. Biologists point out that the Namib beetles have hundreds of tiny bumps on their backs, and those protrusions attract water molecules from fog or dew. In fact, each day that insect is able to collect an amount of water equal to 12 percent of its body weight. Impressed by the creature's water-collection ability, the NBD Nano team is trying to mimic it by developing a bottle with a surface covered by little bumps like those on the beetle's back. They hope that when the new technology is perfected, a single such bottle will be able to collect and store up to 3.2 quarts (3 L) of water per hour.

That peculiar wing arrangement consists of a series of feathers that have rough edges, not unlike the bristles of a stiff hairbrush. These feathers act to break up sound waves generated by the owl's beating wings. Hence, those wings make far less noise than the wings of most other birds, which have smoother edges.

Liu and his fellow researchers believed they could reduce turbine noise by developing blades that mimicked the structural configuration of owl wings. Working with a blade shape known as an airfoil, they added serrated edges, somewhat like the edge of a steak knife. And when they forced a stream of air over the rotating blades, it became immediately clear that the noise reduction was significant.

Owl-inspired research into sound reduction for wind turbines is still in its infancy. And even when the technology is perfected, it will take time to retrofit the many existing turbines with the improved blades. But the scientists at Xi'an Jiaotong University are hopeful. As Alison Bosman, a staff writer for the environmental website Earth.com, says, they "believe their work will serve as an important guide for airfoil design and noise control in the future."[43]

Ideas from a Tiny Creature's Guts

Just as biomimicry based on owls' wings can make operating a wind turbine more energy efficient, imitating a special ability of another creature has made light-emitting diode (LED) lightbulbs more efficient. An LED lightbulb works by passing an electrical current through a device called a diode, thereby making it glow. Even before the biomimetic improvement in LEDs occurred, those devices were about 90 percent more efficient than ordinary lightbulbs.

For a long time, however, there was a noticeable drawback to LEDs. Namely, some of the light in the tube containing the diode reflected back at its source. This made the glow look dimmer than if it were not reflected. Desiring to make LED lights more efficient and thereby brighter, in the early 2000s Belgian physicist Jean-Pol Vigneron studied the light-producing capabilities of fireflies, also called lightning bugs. Those unusual insects create light via a chemical reaction within their abdomens. That glow shines through their body's outer layer, called the cuticle. Using a high-powered microscope, Vigneron saw that the cuticle is studded with many tiny triangular scales that protrude at varied angles. These jagged structures reflect the light from the abdomen back and forth, in effect making the glow look brighter.

Copying that process, Vigneron and his team of researchers used a laser beam to cut a tiny pattern of triangular scales into some light-sensitive material. They then attached those scales to the outer surface of a standard LED bulb. This caused much of the light from the diode to reflect outward rather than back toward the source, thereby making the LED at least 50 percent brighter.

Penguins, Bubbles, and Energy Efficiency

Just as it has helped make LED lightbulbs, wind turbines, and solar cells more efficient, biomimicry has been employed to increase the efficiency of engines fueled by oil and gasoline. One recent, well-publicized example involves tanker and cargo ships and other large vessels. To move through the water quickly and efficiently, such ships must use a lot of heavy fuel oil, which is expensive. Also, when it is burned, it emits a good deal of the greenhouse gas CO_2, which contributes to climate change by making the atmosphere warmer.

Desiring to find a way for big ships to use less fuel and thereby both save money and reduce CO_2 emissions, in the early 2000s scientists in multiple countries turned to biomimicry. The specific animals they chose to imitate were emperor penguins. Emperor penguins have the uncanny ability to speedily move upward through the water and suddenly launch themselves into the air like small-scale rockets. A penguin performs this maneuver by first swimming to the surface and there trapping air between its body and wings. Next it dives downward, rapidly spins around, and just as quickly moves upward again. During that ascent, the bird releases the air from beneath its wings, forming an envelope of bubbles. That layer of air reduces drag on the creature's body, enabling it to achieve rocket-like propulsion as it clears the surface.

An emperor penguin achieves rocket-like propulsion as it clears the water's surface. If scientists can replicate this motion, they might be able to develop ships that require less fuel and produce fewer harmful emissions.

"Biomimicry breakthroughs in energy efficiency are just a few of the ways emulating nature is shaping the future of sustainable innovation."[44]

—Virginia Energy Efficiency Council

The researchers reasoned that creating the same sort of layer of air bubbles around a ship's hull might reduce drag in a similar manner. To that end, they created a device that came to be called the Air Lubrication System, or ALS. It injected air into the water beside and beneath a ship, forming the desired layer of bubbles. And sure enough, initial tests of the system showed that drag was indeed reduced, allowing the vessel to go somewhat faster while using significantly less fuel. Furthermore, there was at least a 10 percent reduction in CO_2 emissions during the tests. Although not all large vessels have adopted the ALS, it is swiftly growing in popularity, and its use is expected to eventually become a shipping industry standard.

Penguins, fireflies, owls, and whales are only a few of the animals scientists are currently investigating in hopes of conserving energy. According to the Virginia Energy Efficiency Council, a nonprofit group that advises the state government on ways to save energy, "Reducing energy use is one problem scientists and engineers are tackling by looking at the world around them to inspire solutions. These biomimicry breakthroughs in energy efficiency are just a few of the ways emulating nature is shaping the future of sustainable innovation."[44]

CHAPTER FIVE

The Realm of Robotics

In February 2023 a team of Canadian researchers demonstrated the abilities of a new robot they call GeiwBot. The name is a mash-up of the two creatures that inspired its development: the gecko and the inchworm. GeiwBot is fairly tiny—only about 1.6 inches (4 cm) long, 0.12 inch (3 mm) wide, and 0.04 inch (1 mm) thick. Like the inchworm, the bot moves forward by repeatedly arching, stretching, and temporarily deforming the shape of its body. At the same time, like the gecko, the bot has adhesive properties in its footpads. These allow it to sturdily attach itself to almost any surface and thereby to walk up and down walls and even across ceilings.

Although GeiwBot is not yet being widely used in real life, researchers envision several possible scenarios for which a tiny, mobile, highly responsive device could be useful. One of the most promising potential uses for it is in search and rescue operations following building collapses. The biggest difficulty in such operations, explains Robin R. Murphy, an expert on robot-assisted rescue at Texas A&M University, "is seeing inside the rubble." Telling an interviewer about how a robot like GeiwBot might find victims trapped in such rubble, she continues:

> You've got basically a concrete, sheetrock, piping and furniture version of pickup sticks. If you can get a robot into the rubble, then the structural engineers can see the interior of that pile of pickup sticks and say "Oh, OK, we're not going [to] pull on that, that's going to cause a secondary collapse. OK, we should start on this side, we'll get through the debris quicker and safer." Going inside rubble piles is really hard. Scale is important. If the void spaces are on the order of the size of the robot, it's tricky.[45]

A device like GeiwBot may be ideal for such chaotic situations, Murphy says. This is partly because it is strong and partly because it is capable of climbing through the remains of a collapsed structure at many different angles. In addition, it is able to cling to a wide array of materials. Once it is equipped with a tiny camera, she asserts, such a robot will likely end up saving many lives.

Canadian researchers have developed a tiny, mobile robot that could be used to find people trapped in collapsed buildings. The robot is a mash-up of a gecko, which can climb walls thanks to its sticky feet, and an inchworm, which arches and stretches to move forward.

An Explosion of Technical Advances

Murphy's optimism over the success of some preliminary tests of GeiwBot's abilities is mirrored by similar feelings about the modern robotics industry among engineers and designers within that scientific discipline. This is because of a genuine explosion of technical advances made in that field in the past three decades. Importantly, states technology reporter Melissa Bradshaw, a large proportion of those strides in the realm of robotics have involved biomimicry. "Robotic advances inspired by the natural world," she explains,

> are providing solutions to some of our most pressing environmental challenges. . . . As we look to a future in which advanced robotics technologies are ever-more integrated with our everyday lives, the idea that robots might help us protect the planet we live on is a particularly exciting prospect. While "robots saving the Earth" sounds slightly farfetched, we're now edging towards a reality in which machines could provide the assistance we need to solve a range of critical environmental challenges: challenges that humans alone aren't equipped to solve.[46]

The world's many different animal species—each with their own unique characteristics and behaviors—present a seemingly endless variety of models for such robotic inventions. Mark Berezniak, a science writer who specializes in reporting on the robotics industry, notes, "[Some] engineers and scientists have found nature as a promising inspiration for . . . design, innovation, and perfection. They have analyzed and studied the animal kingdom to figure out how their body, skeletal, and muscle mechanism works to create . . . [more] efficient robots and machines."[47]

"Robotic advances inspired by the natural world are providing solutions to some of our most pressing environmental challenges."[46]

—Technology reporter Melissa Bradshaw

The Robird (pictured) is a robot drone that looks and acts like a real bird. Because it is so lifelike, it is capable of scaring birds away from airports, where they often pose a danger by flying into airplane engines.

One of several examples Berezniak has reported on in recent years is the Robird—a bird-shaped robot drone with flapping wings. Developed by the Dutch company Clear Flight Solutions, the Robird was designed to scare real birds away from airports, where they often pose a danger by flying into airplane engines. The magic of the Robird is its similarity to a real bird. As Wessel Straatman, a chief research and development scientist for the company, explains:

> By mimicking their natural counterparts through silhouette and behavior, they are indistinguishable from real-life birds of prey to other birds, and circling airports will be enough to stop real birds from straying into airplane flight paths. These remote-controlled robotic birds, with the realistic appearance and weight of their living counterparts, serve as modern-day scarecrows.[48]

Robots in Orbit and Beyond

The animal kingdom is even influencing innovations intended for use in space. One example, developed by a team of Stanford University scientists in conjunction with the National Aeronautics and Space Administration (NASA), is the Stickybot. As its name suggests, its design was partly inspired by the sticky adhesive pads found on gecko feet. Of the two principal uses for Stickybot, as conceived by its inventors, the first is to walk along the outer surface of an orbiting spacecraft or space station. Having reached a preplanned spot, the bot will attach itself firmly with its gecko-like feet. At that point, it can either take photos of something that needs to be repaired or actually do some of those repairs.

The second main job envisioned for Stickybot is to become, in effect, the outer-space version of a city street cleaner. Berezniak explains that "around 500,000 pieces of manmade space junk orbits Earth at frankly terrifying speeds. The aim of this cleanup gripper is to [be] able to grasp large objects, weighing as much as 370kg, and remove them."[49] Thus, several orbiting Stickybots would collect the debris and deposit it in some sort of orbiting storage facility.

Meanwhile, another team of NASA researchers has been working on a different kind of biomimetic space robot. Far beyond the region of near-Earth orbit loom two celestial orbs—the moon and Mars—to which NASA and other space agencies around the world have been sending mechanical probes and landers for several decades. One problem experienced during those expeditions has been that the landers, or rovers, are too big to fit into most of the small crevices they encounter on those worlds.

> **"A Snakebot could navigate over rough, steep terrain where a wheeled robotic rover would likely get stuck or topple."[50]**
>
> —NASA researcher Gary Haith

To solve that problem, NASA has developed Snakebots. As the name implies, they are robots shaped like snakes. In the words of Gary Haith, one of the NASA researchers who is developing the bots, "A Snakebot could navigate over rough, steep terrain

Flying Mechanical Pollinators

Bees have long been one of nature's most industrious and important animal species. This is because they transfer pollen from one plant to another. That process of pollination is crucial to almost all the crops that humans grow around the world. An increasing problem, however, is that a combination of climate change, pesticide use, insect diseases, and other detrimental factors has caused major decreases in bee populations. Hoping to help alleviate this problem, in the 2010s researchers at Harvard's School of Engineering and Applied Sciences developed what they dubbed RoboBee, a robot based on the bumblebee. It is about the size of a quarter, and its wings flap at a rate of 120 times per second, allowing it to fly, like a tiny drone, from garden to garden and field to field. A pollination and docking appendage in the bot's backside can carry and deliver pollen to a flower very much like a bee does.

where a wheeled robotic rover would likely get stuck or topple."[50] An example of the bot's use, Haith continues, would be if such a wheeled rover rolls across the Martian surface and encounters a crevice. Mission controllers back on Earth decide they want to explore that fissure, so they send a signal to the rover. The signal commands the craft to remove a Snakebot from an onboard storage chamber and place it on the ground near the crevice. The bot will then receive a signal ordering it to enter the crevice and snap photos, dig into the soil, or both.

Underwater Helpers

As it turns out, the form and flexibility of snake bodies make robots shaped like them useful for a good deal more than exploring tight spaces on other worlds. Snakebots are also ideal for doing various tasks in earthly environments that are difficult for people to work in. One of those settings consists of the undersides of ships and submarines at sea. Now and then something goes wrong when such a vessel is far from port facilities. If the crew needs to inspect the ship's bottom right away, they must send down specially trained divers, who may not be available. Even if they *are* available, a storm or some other adverse condition may make it too dangerous for them to operate.

Another possibility is to dispatch a submersible that can hold one or two people. But such devices are extremely expensive and require specialists to operate them. So only a few ships carry them.

Far cheaper would be to use an underwater Snakebot that can be operated by radio signals, like a drone. The company Advanced Robotics for Manufacturing (ARM), located in Pittsburgh, Pennsylvania, developed just such a bot during 2017 to 2019. Such a device can allow a crew to obtain data about a mechanical problem well before the ship reaches its home port, says Matt Fischer, a program manager at ARM. "That saves weeks or months of time in a maintenance schedule," he points out. "And in turn, that saves money."[51]

A different kind of underwater robot, also based on the bodily structure of an animal, is the MantaDroid, created by researchers at the National University of Singapore. In 2017 a team headed by Chew Chee-Meng announced the completion of this submersible that looks almost exactly like a manta ray. Manta rays are

The underwater robot, MantaDroid (pictured), is a submersible that looks and swims like a manta ray. It has potential for use in underwater surveying, search and rescue, and data collection.

Giving Cockroaches a Good Name

Cockroaches are not the most inspirational of creatures. And yet they were the inspiration behind a robot designed to crawl through small spaces found in the rubble of collapsed structures in search of survivors. During the 2010s a team of researchers at the University of California, Berkeley, invented a small, six-legged robot that looks somewhat like a cockroach and moves remarkably like that insect. The team members named it the Dynamic Autonomous Sprawled Hexapod, or DASH for short. Made of cardboard and old recycled electronic parts, the bot is inexpensive to make. The legs are powered by a small, relatively cheap direct current motor, a device that converts electrical energy into mechanical energy. Like real cockroaches, when DASH moves head-on into a barrier, it scurries upward, trying to clear it. And if it encounters an open space above the barrier, it propels itself into that new space. A tiny built-in camera takes photos as DASH moves along. Some of the bot's makers have joked that the little bot gives lowly cockroaches a good name.

among the finest swimmers in the seas. An adult of the species can move along at 9 miles per hour (14.5 kph), almost twice as fast as an Olympic swimmer. Appendages, called wings, extend outward from the creature's sides and flap almost like bird wings to smoothly propel the ray through water.

For years scientists had tried unsuccessfully to imitate that motion in a robot. Chee-Meng's team managed to replicate this motion in an amazingly realistic fashion, in part by placing a small but complex motor in each wing. As a result, MantaDroid can swim 2.3 feet (70 cm) per second for up to ten hours. Predicting how this ability will allow the bot to perform various tasks, Chee-Meng comments,

> we see MantaDroid as a potential long-term continuous underwater surveying machine, which could be useful for maritime industries, environmental agencies, and search and rescue organizations. For example, it can be used for underwater inspection tasks, as well as for collection of [undersea] data. . . . [Also] multiple MantaDroids [could] be deployed to . . . perform search operations, such as looking for lost divers or sunken objects in the sea.[52]

Work of Vast Scope and Potential

Robots that can crawl through crevices on Mars, chase away birds at airports, and do a multitude of jobs beneath the waves represent only a handful of modern machines that mimic diverse aspects of nature. Each year many more such projects are launched in labs worldwide as scientists take advantage of what nature has learned over the span of countless eons. Thanks in part to the vast scope and potential of this work, Jay Harman remarks,

> this a very exciting time in science and technology. I am inspired daily by the potential of applying nature's lessons to design a new golden age for the Earth and for humanity—a golden age that is not only possible, but realistically achievable. Biomimicry will get us there. Whether you're a [chief executive officer], corporate employee, commercial manufacturer, entrepreneur, politician, small business owner, college student looking to start a company, teacher wanting to share a world of positive choices with your students, or merely someone curious about [biomimicry], my hope is that one message rings loud and clear: By learning from nature, we can create more abundant, healthy, satisfying lives for ourselves, our children, and our planet.[53]

SOURCE NOTES

Introduction: Attempts to Copy Nature's Designs

1. Jeremy Williams, "Building of the Week: Eastgate, Zimbabwe," *The Earthbound Report* (blog), May 15, 2020. https://earthbound.report.
2. Biomimicry Institute, "Biomimicry Is a Practice That Learns from and Mimics the Strategies Found in Nature to Solve Human Design Challenges—and Find Hope," 2023. https://biomimicry.org.
3. Biomimicry Institute, "Biomimicry Is a Practice That Learns from and Mimics the Strategies Found in Nature to Solve Human Design Challenges—and Find Hope."
4. Christopher J. Botham, "Leonardo da Vinci and Human Flight," *On Verticality* (blog), April 9, 2022. www.onverticality.com.
5. Biomimicry Institute, "Biomimicry Is a Practice That Learns from and Mimics the Strategies Found in Nature to Solve Human Design Challenges—and Find Hope."

Chapter One: Engineering and Construction

6. Janine M. Benyus, *Biomimicry: Innovation Inspired by Nature*. New York: HarperCollins, 2002, p. 96.
7. Eszter Gurbicz, "BIOHM: The New Sustainability Standard in Construction Industry," Staiy, October 10, 2022. https://staiy.com.
8. Gurbicz, "BIOHM."
9. Jay Harman, *The Shark's Paintbrush: Biomimicry and How Nature Is Inspiring Innovation*. Ashland, OR: White Cloud, 2013, p. 27.
10. Kristy Hamilton, *Nature's Wild Ideas: How the Natural World Is Inspiring Innovation*. Vancouver, BC: Greystone, 2022, p. 132.
11. Nina Tory-Henderson, "The Hive Learning Hub," Danish Architecture Center. https://dac.dk.
12. Tory-Henderson, "The Hive Learning Hub."
13. Eda N. Dalgiç, "Biomimicry: Inspiring Nature for Architecture," Archeetect, September 24, 2020. https://archeetect.com.
14. Dalgiç, "Biomimicry."
15. Lawson Brandis, "Biomimicry Part 2—the Japanese Bullet Train," Integrate Sustainability, January 11, 2018. www.integratesustainability.com.au.
16. Donovan Alexander, "Biomimicry: 9 Ways Engineers Have Been 'Inspired' by Nature," Interesting Engineering, November 28, 2018. https://interestingengineering.com.

Chapter Two: Medical Applications

17. Quoted in Julia Belluz, "How a Lizard's Venom Inspired the Promising Weight Loss Drug Wegovy," Vox, July 5, 2021. www.vox.com.
18. Quoted in Belluz, "How a Lizard's Venom Inspired the Promising Weight Loss Drug Wegovy."
19. Belluz, "How a Lizard's Venom Inspired the Promising Weight Loss Drug Wegovy."
20. Juli Berwald, "The Immortal Jellyfish," *Discover*, November 9, 2017. www.discovermagazine.com.
21. Amber Dance, "Some Salamanders Can Regrow Lost Body Parts. Could Humans One Day Do the Same?," *Smithsonian*, January 20, 2020. www.smithsonianmag.com.
22. Dance, "Some Salamanders Can Regrow Lost Body Parts."
23. Josh Cassidy and Laura Shields, "Porcupine Barbs for Better Wound Healing," NPR, April 9, 2019. www.npr.org.
24. Quoted in Cassidy and Shields, "Porcupine Barbs for Better Wound Healing."
25. Harman, *The Shark's Paintbrush*, p. 2.
26. Quoted in Dance, "Some Salamanders Can Regrow Lost Body Parts."

Chapter Three: Protective Gear for Soldiers and Athletes

27. David Szondy, "Cephalopods Could Be the Key to Better Camouflage," New Atlas, February 26, 2018. https://newatlas.com.
28. Szondy, "Cephalopods Could Be the Key to Better Camouflage."
29. Quoted in Aaron Gouveia, "Turtle Research May Aid Military Armor," *Cape Cod Times* (Hyannis, MA), February 26, 2009. www.capecodtimes.com.
30. Quoted in Gouveia, "Turtle Research May Aid Military Armor."
31. Quoted in Stacey Liberatore, "Forget the Wetsuit, Scientists Say Donning a Fur-Like 'Beaversuit' Could Keep You Warm in the Water," *Daily Mail* (London), October 5, 2016. www.dailymail.co.uk.
32. Harman, *The Shark's Paintbrush*, p. 87.
33. Harman, *The Shark's Paintbrush*, p. 87.
34. Quoted in Doug Samuels, "A Re-Designed Helmet from Mississippi State Researchers Claims to Be 3x More Likely to Reduce Concussions," FootballScoop, March 23, 2017. https://footballscoop.com.
35. Quoted in Viviane Callier, "Head-Banging Woodpeckers Could Give Themselves a Concussion Every Day: Here's How They Avoid It," *Scientific American*, July 14, 2022. www.scientificamerican.com.

Chapter Four: Energy

36. Molly Liempriere, "Could Biomimicry Revolutionise Renewable Energy?," Power Technology, April 26, 2017. www.power-technology.com.

37. Quoted in Mark Shwartz, "Insect Eyes Inspire New Solar Cell Design by Stanford Researchers," Stanford News, August 31, 2017. https://news.stanford.edu.
38. Quoted in Shwartz, "Insect Eyes Inspire New Solar Cell Design by Stanford Researchers."
39. Quoted in David L. Chandler, "Explained: Why Perovskites Could Take Solar Cells to New Heights," MIT News, July 15, 2022. https://news.mit.edu.
40. Harman, *The Shark's Paintbrush*, p. 99.
41. Harman, *The Shark's Paintbrush*, p. 99.
42. Quoted in Alison Bosman, "Owl Wings Inspire Noise Reduction in Wind Turbines and Aircraft," Earth.com, January 18, 2022. www.earth.com.
43. Quoted in Bosman, "Owl Wings Inspire Noise Reduction in Wind Turbines and Aircraft."
44. Virginia Energy Efficiency Council, "Biomimicry and Energy Efficiency," April 19, 2018. https://vaeec.org.

Chapter Five: The Realm of Robotics

45. Quoted in The Conversation, "An Expert on Search and Rescue Robots Explains the Technologies Used in Disasters Like the Florida Condo Collapse," June 30, 2021. https://theconversation.com.
46. Melissa Bradshaw, "Inspired by Nature: Could Biomimicry Robots Help Save the Planet?," *The Engineer*, May 16, 2022. www.theengineer.co.uk.
47. Mark Berezniak, "10 Best Robots Inspired by Nature—Biomimicry," Explore Biotech, November 25, 2018. https://explorebiotech.com.
48. Quoted in Berezniak, "10 Best Robots Inspired by Nature—Biomimicry."
49. Berezniak, "10 Best Robots Inspired by Nature—Biomimicry."
50. Quoted in Berezniak, "10 Best Robots Inspired by Nature—Biomimicry."
51. Quoted in Aaron Aupperlee, "CMU's Snakebot Goes for a Swim," Carnegie Mellon University, April 13, 2021. www.cmu.edu.
52. Quoted in Luke Dormehl, "MantaDroid Robot May One Day Join Its Inspiration in the Depths of the Ocean," Digital Trends, November 9, 2017. www.digitaltrends.com.
53. Harman, *The Shark's Paintbrush*, p. 9.

FOR FURTHER RESEARCH

Books

Karen Ansberry, *Nature Did It First*. Nevada City, CA: Dawn, 2020.

Megan Clendenan and Kim R. Woolcock, *Design Like Nature*. Victoria, BC (Canada): Orca, 2021.

Henry Dicks, *The Biomimicry Revolution: Learning from Nature How to Inhabit Earth*. New York: Columbia University Press, 2023.

Christiane Dorion, *Invented by Nature: Meet the Creatures Who Inspired Our Everyday Technology*. Minneapolis: Quarto, 2021.

Kristen Nordstrom, *Mimic Makers*. Watertown, MA: Charlesbridge, 2021.

Internet Sources

APlanet, "Biomimicry: 10 Creations Inspired by Nature," April 21, 2022. https://aplanet.org.

Ahamefula Ascension, "10 Awesome Examples of Biomimicry for Kids and Scholars," Environment Go!, November 23, 2022. https://environmentgo.com.

Eda N. Dalgiç, "Biomimicry: Inspiring Nature for Architecture," Archeetect, September 24, 2020. https://archeetect.com.

FibreGuard, "Biomimicry in Architecture: How Designers Look to Nature," December 9, 2022. https://fibreguard.com.

Indeed Editorial Team, "What Is Biomimicry Engineering? (With Examples and Jobs)," Indeed, September 30, 2022. www.indeed.com.

Just Style, "Biomimicry—a Key Solution for Fashion Sustainability," 2023. www.just-style.com.

Learn Biomimicry, "The Top Biomimicry Courses and Certificates of 2023," February 27, 2023. www.learnbiomimicry.com.

Learn Biomimicry, "The Top 10 Examples of Biomimicry in Architecture and the Built Environment," February 11, 2023. www.learnbiomimicry.com.

Eddie Phillips, "Biomimicry Is Real World Inspiration," Carnegie Museum of Natural History, 2020. https://carnegiemnh.org.

Sasha Weilbaker, "10 Modern Inventions Modeled After Nature," *Mental Floss*, July 6, 2022. www.mentalfloss.com.

Jeremy Williams, "Building of the Week: Eastgate, Zimbabwe," *The Earthbound Report* (blog), May 15, 2020. https://earthbound.report.

Organizations and Websites

Biomimicry Institute

https://biomimicry.org

The Biomimicry Institute says that its mission is to help solve many of the major challenges humanity faces by utilizing designs from nature. The website features ways that young people can become involved, including contests in which students can submit their own biomimicry ideas.

ECOncrete

https://econcretetech.com

A company founded by marine biologists, ECOncrete creates strong, ocean-friendly marine infrastructure. This site explains the benefits of biomimicry and provides several links to ongoing programs in which students and other amateurs can submit their own biomimicry designs.

University of Akron Biomimicry Research and Innovation Center

www.uakron.edu/bric

The University of Akron Biomimicry Research and Innovation Center is a world leader in biomimicry research, with extensive courses and lab studies related to biomimicry. In addition, the center offers a wide array of educational opportunities, including student design challenges and fellowships.

INDEX

Note: Boldface page numbers indicate illustrations.

PICTURE CREDITS

Cover: Shutterstock.com

7: Raj Singh/Alamy Stock Photo
10: Suzanne Tucker/Shutterstock; Josep Curto/Shutterstock (inset)
12: robertharding/Alamy Stock Photo
15: IR Stone/Shutterstock
20: Erin Donalson/Shutterstock.com
22: sportpoint/Alamy Stock Photo
24: Gerald Corsi/iStock
28: RayK Photos/Shutterstock
31: Simone Hogan/Alamy Stock Photo
35: Sean R. Stubben/Shutterstock
38: Tomas Rak/Alamy Stock Photo
41: Evelyn D. Harrison/Shutterstock
44: Iurii Kazakov/Shutterstock
47: SORNNARIN NANKRATOK/Shutterstock; Chantelle Bosch/ Shutterstock (inset bottom); DaydreamsGirl/iStock (top inset)
49: ANP/Newscom
52: Ferrari/ZUMA Press/Newscom

ABOUT THE AUTHOR

Classical historian, amateur astronomer, and award-winning author Don Nardo has written numerous volumes about scientific topics, including *Destined for Space* (winner of the Eugene M. Emme Award for best astronomical literature); *Tycho Brahe* (winner of the National Science Teaching Association's best book of the year); *Deadliest Dinosaurs*; and *The History of Science*. Nardo, who also composes and arranges orchestral music, lives with his wife, Christine, in Massachusetts.